Health 241

永葆生机
Never Too Old

Gunter Pauli

[比] 冈特·鲍利 著

[哥伦] 凯瑟琳娜·巴赫 绘

章里西 译

上海远东出版社

丛书编委会

主　任: 贾　峰

副主任: 何家振　闫世东　郑立明

委　员: 李原原　祝真旭　牛玲娟　梁雅丽　任泽林

　　　　王　岢　陈　卫　郑循如　吴建民　彭　勇

　　　　王梦雨　戴　虹　靳增江　孟　蝶　崔晓晓

特别感谢以下热心人士对童书工作的支持:

匡志强　方　芳　宋小华　解　东　厉　云　李　婧

刘　丹　熊彩虹　罗淑怡　旷　婉　杨　荣　刘学振

何圣霖　王必斗　潘林平　熊志强　廖清州　谭燕宁

王　征　白　纯　张林霞　寿颖慧　罗　佳　傅　俊

胡海朋　白永喆　韦小宏　李　杰　欧　亮

目录

Contents

一棵橡树正在为即将到来的冬天做准备。一棵柳树看到橡树的叶子在变色，于是问道：

"你准备好过冬了吗？"

"哦，经历了400个这样的冬天之后，我已经习惯面对刺骨的寒风和冰冷的天气了。"

An oak tree is getting ready for the coming winter. A willow observes how the oak's leaves are changing colour and asks her:

"Did you prepare well for the cold winter?"

"Oh, getting ready to face the harsh winds and the freezing cold has become routine for me, after having done so four hundred times before."

一棵橡树正在为即将到来的冬天做准备……

An oak tree is getting ready for winter ...

有些橡树已经有1000年的树龄了……

Oak trees that are a thousand years old ...

"别告诉我你已经强壮地活400年了。这太惊人了！"

"实际上，我一点也不特别。周围有些橡树已经有1000年的树龄了，这一点也不稀奇。"

"告诉我，你们橡树是怎么知道冬天要来临的？在夏天遇到寒流时又是怎么避免落叶的？"

"Don't tell me you've been going strong for four hundred years. That's astounding!"

"I am not at all unique, you know. There are oak trees around that are a thousand years old, and not even that is exceptional."

"Tell me, how do you oaks know when winter is really approaching, and how do you avoid dropping your leaves during a cold spell in summer?"

"哦，天冷的时候我能看到。"

"没人能看到天气变冷。我知道你聪明，但是，请你千万不要认为我会蠢到相信你能看到寒冷！"

"无意冒犯，小柳树，但这是真的。我能看到太阳发出的红光，当红光变少时，就意味着冬天真的要来了。"

"Well, I can see when it's getting cold."

"No one can see it's getting cold. I know you are wise but please, Oak, surely you don't consider me stupid enough to believe that you see it is cold!"

"No offence, young Willow, but it is true. I am able to see the red light of the sun, and when there is less of it, it means winter is really setting in."

我能看到太阳发出的红光。

I can see the red light of the sun.

海豚用鼻子"看"……

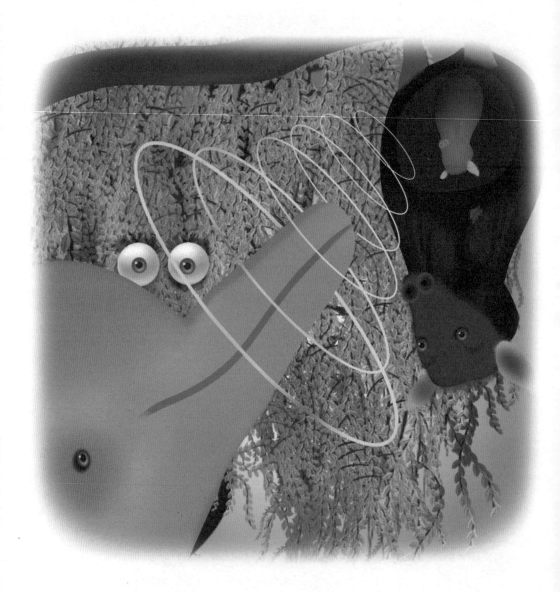

A dolphin "sees" with its nose…

"你真的能看到吗？我们这些树都没有眼睛，你是怎么做到的呢？"

"谁说我们需要眼睛才能看见？海豚用鼻子'看'，蝙蝠用耳朵'看'。你猜怎么着？我们树也会说话。大自然不是充满了惊喜吗？"

"You actually see it? How could you, when we trees don't have eyes?"

"Who says we need eyes to see? A dolphin 'sees' with its nose, and a bat with its ears. And guess what? We trees can talk as well. Isn't Nature just full of surprises!"

"橡树，你真的把我搞得晕头转向了。你说你会说话，可你连嘴都没有！"

"我在跟你说话，是不是？我也自言自语。只听自己已经熟知的东西多无聊。生活充满了惊喜。"

"Now you're really pushing me beyond my comfort zone, Oak. You claim youcan talk – but you don't even have a mouth!"

"I am talking to you, aren't I? And I talk to myself too. How boring, to only listen to what you already know. Life is full of surprises."

我在跟你说话，是不是？

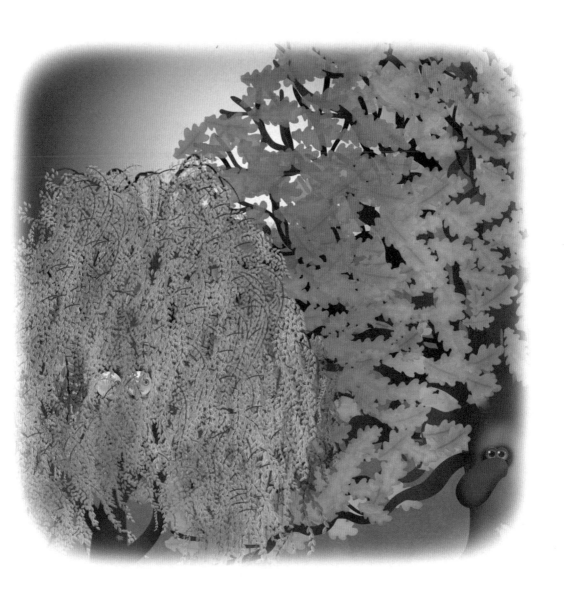

I am talking to you, aren't I ?

让它们的叶液尝起来很难吃……

Make the sap of their leaves taste bad ...

"所以你会对着自己的枝条喃喃自语，"柳树笑着说，"告诉我，你都说了些什么？"

"看，当我感觉到一只毛毛虫在吃一根树枝上的叶子时，我就会马上告诉其他树枝，让它们的叶液尝起来很难吃，这样虫子就不会去吃其他叶子了。"

"So you mumble to your branches," Willow laughs. "And do tell me, what is it you talk to yourself about?"

"Look, when I feel a caterpillar eating the leaves on one of my branches, I quickly tell the other branches to make the sap of their leaves taste bad, so insects will leave the rest of the tree alone."

"可是为什么毛毛虫就不能吃一片你的叶子呢？有人告诉我，你可是有多达100万片叶子呢。这是真的吗？"

"100万有点夸张了，不过，我可能有70多万片叶子。但是，这里有几百种昆虫，每一种都有成千上万个家庭成员……我必须一个接一个地对付它们。"

"你知道是谁在享用你的叶子吗？"

"But why can't a caterpillar have one of your leaves? Of which I am told you may have as many as a million. Is this true?"

"One million is a bit on the high side, but yes, I may have well over 700,000 leaves. But then, there are hundreds of insects, and each has thousands of family members… And I have to deal with each of them, one by one."

"Do you know who it is feasting on you?"

70多万片叶子

Over 700,000 leaves

......在血管里注射一些防冻剂......

...some anti-freeze into my veins ...

"不知道，我分辨不出。但我确实能感觉到它们，然后用一杯鸡尾酒来回应它们，让它们的好胃口消失。"

"你真叫人吃惊！"

"哦，够了，别再恭维我了！对不起，我得忙了，在冬天真正来临之前，我需要在血管里注射一些防冻剂。"

"什么！防冻剂？"

"No, that I cannot see. But I do sense them, and then respond with a cocktail that will kill their appetite, and only theirs."

"You are indeed astounding!"

"Oh, enough with the flattery! Excuse me, I have to get on now, and put some anti-freeze into my veins, before the winter really hits."

"Come on! Anti-freeze?"

"对的，我要把树液中的水挤出来，只剩下糖，这样我的木头就像冰一样不会开裂了，我就能挨过严寒，在明年秋天再结出成千上万颗橡子。你看，我们长久生存的核心在于规划。"

"所以明年冬天松鼠们也能有足够的储粮了，真是周密的规划。真的，任何生灵都有机会永葆生机。"

……这仅仅是开始！……

"Yip, when I squeeze out the water from my sap, only sugars are left. Now my wood won't crack, like ice does in the freezing weather. And I'll survive, to produce thousands of acorns next fall. We need to plan, you see."

"So the squirrels too will have plenty to eat next winter. Good planning for the years to come. Indeed, we are never too old for anything!"

... AND IT HAS ONLY JUST BEGUN!...

... AND IT HAS ONLY JUST BEGUN! ...

Oak trees are strong and inflexible; as a result their branches will break off in storms. Willows are flexible and their branches move with the force of the wind, so they hardly ever lose one through force.

橡树粗壮笔直，所以树枝容易在暴风雨中折断。柳树柔韧性强，枝条可以随风摆动，所以不太会因为大风而折断。

250 000 升
CO_2

An oak tree releases 250,000 litres of oxygen per year. A hundred-year-old oak weighs ten tons, and will add 250 kg of wood every year. While it is growing leaves in the spring, it can extract water from under the ground at a rate of up to 70 litres per hour.

一棵橡树每年释放 25 万升氧气。一棵百年橡树重达 10 吨，每年会增加 250 千克木材。当橡树春天长叶子时，它能以每小时 70 升的速度从地下吸取水分。

Oak trees have photosensors that can detect red light from the sun. This enables the tree to grow towards light, and to be aware of the shortening or lengthening of the day, signalling the advent of different seasons. These sensors also act as temperature sensors.

橡树有感光器，可以探测到来自太阳的红光。这使得橡树向光生长，并能意识到白天的缩短或延长，而这标志着不同季节的到来。这些传感器还充当温度传感器。

When celebrating a wedding anniversary, a 25th anniversary is silver, a 50th gold, a 60th platinum, and a 70th a diamond anniversary. An 80th wedding anniversary, which is reached very seldom, is an "oak" anniversary.

庆祝结婚的纪念日中，25周年是银婚，50周年是金婚，60周年是铂金婚，70周年是钻石婚。结婚80周年是"橡树"婚，这是很少能够达到的。

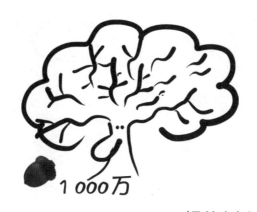

1 000万

A single oak tree can produce up to about 10 million acorns during its lifetime. In one fall it could produce 2,000, and the next 25,000. This sheer abundance ensures more acorns than can be eaten by animals.

一棵橡树在其一生中可以产出大约1 000万颗橡子。若某个秋天它能产出2 000颗，第二年秋天就能产出25 000颗。这种绝对的高产确保了动物们可以吃到更多的橡子。

An oak gall is an acorn deformed by a wasp, and contains tannic acid which, when mixed with iron sulphate (FeSO$_4$), creates an ink. This is the type of ink used by Da Vinci for his drawings. It is also the type of ink in which the Magna Carta, the US Declaration of Independence, was written.

橡树瘿是一种由黄蜂引发异常生长的橡子，含有鞣酸，当与硫酸铁混合时可以制成墨水。达芬奇就是用这种墨水来作画的。这也是拟定《大宪章》和《美国独立宣言》时所用的墨水。

The permanence and water resistance of gallnut ink made it the standard writing ink for drip pens in Europe for over 14 centuries. It is not suitable for use in fountain pens that operate on the principle of capillary forces.

鞣酸墨水的耐用性和耐水性使其成为欧洲 14 个世纪以来的标准书写墨水。它不适合用于钢笔，钢笔是根据毛细管原理来书写的。

Due to its tannin content, the leaves and acorns of oak trees are toxic for horses, cattle, sheep and goats. Acorns do, however, form part of the diet of birds, pigs, deer, squirrels, and mice. Acorns leached of tannins are good for human consumption.

橡树叶子和橡子含有单宁，马、牛、绵羊和山羊食用后会中毒。然而，橡子却是鸟类、猪、鹿、松鼠和老鼠的食物来源之一。含有单宁的橡子对人体有益。

What do you think of living for a thousand years?

你觉得活1000岁会怎样?

Can you sense the sun with your skin?

你的皮肤能感觉到阳光吗?

How is it that acorns are toxic for some but not other animals?

为什么橡子对某些动物有毒而对其他动物无害呢?

Would you like to write with ink that cannot be erased?

你愿意用擦不掉的墨水写字吗?

How many oak trees are around where you live? Go and have a close look at one. First of all, touch the bark and feel its texture. Now look for other life forms on the tree. You will quickly find that the tree is like a zoo, with dozens of other living creatures. Make a list of every species you can spot, even the smallest ones. Estimate how tall the tree is. How old do you think it is? Is it healthy, or is it suffering from damage by insects or fungi?Do you think this tree will live to a thousand years, or is it likely to die before its natural life span? Write a report on the tree, and make suggestions as to what needs to be done, or not be done, to create healthier living conditions for the tree.

你家附近有多少橡树？找一棵仔细观察。先摸树皮，感受它的质地。然后寻找树上的其他生命。很快你会发现这棵树就像一个动物园，其中生活着许多生物。列出所发现的每个物种，包括非常小的。估计这棵树有多高。你觉得它有多少岁？它是健康的，还是在遭受虫害？你认为这棵树能活1000年，还是会提前死去？写一篇观察报告，并提出相关建议，为这棵树创造更健康的生存环境。

学科知识
Academic Knowledge

生物学	橡树有600多种；从地中海到亚热带雨林，橡树都是关键物种；橡木与真菌有共生关系，包括佩里格德松露和皮埃蒙特白松露；欧洲的花蝇依靠橡树生存；"橡树突然死亡"是由一种能在几周内杀死橡树的霉菌所致；"橡木苹果"是瘿蜂制造的虫瘿；啄木鸟会把橡子埋起来，而且经常会忘记埋在哪里，这样有助于橡子的传播；橡树是280种不同昆虫的寄主；"槲寄生"寄生在橡树上。
化 学	橡木单宁对一些生物有毒，对另一些则有营养；单宁能保护橡树免受昆虫和真菌的伤害；橡树墨水变黑的原因是将铁离子从Fe^{2+}氧化为Fe^{3+}；橡木桶有助于发酵饮料的颜色、香气和口感；橡木片使烟熏奶酪和鱼更有味道。
物 理	钢笔中的墨水是在毛细管作用下用于书写的；橡木单宁是水溶性的，易于让墨水渗入纸张，使墨水不褪色，所以擦除非常困难；橡木的密度很高（75克/立方厘米）。
工程学	橡木木纹为支撑重量和压力提供了巨大的强度，因此在被钢板取代之前被用于制作战舰的内壳；建于13世纪的英国索尔兹伯里大教堂的橡木屋顶和塔楼历经700年依然完好无损，为保证质量，教堂采用了同一时期种植的橡树木。
经济学	橡树有许多用途，特别是作为建筑材料；西班牙的Bellota火腿用橡果饲养的猪肉制作；橡树皮被用来做酒瓶的软木塞；橡木用于制作高质量的鼓和桶。
伦理学	只看到我们已知的现实，会导致否定其他现实的存在——仅仅因为它们不符合我们预先建立的知识。
历 史	希腊神话：宙斯认为橡树是神圣的；9世纪由橡木制成的维京海盗船；凯尔特人认为寄生在橡树上的槲寄生是男性生育能力的象征。
地 理	橡树只在北半球自然生长；在英国，橡树比其他任何林地的树木都多。
数 学	从数一根细枝的叶子开始数树叶，再数一根树枝有多少细枝，然后数一棵树上有多少树枝，选取15个不同大小的叶片，取其平均值计算出树的叶片的总面积；叶子的形状在反射、平移和旋转方面表现出对称。
生活方式	橡子是治疗腹泻和肾结石的天然药物。
社会学	携带橡子是为了祈求好运。
心理学	橡树是力量和耐力的象征。
系统论	墨西哥、中美洲和北安第斯山脉的橡树林已经被砍伐，用于养牛和种植咖啡；78种野生橡树正面临灭绝的危险；橡树为几百个物种提供了生存环境，一些物种威胁到橡树的生存，另一些物种则为橡树提供了大量的营养和生存条件。

情感智慧
Emotional Intelligence

柳 树

柳树表达了他对森林中其他生物的关心。他很难相信橡树的年龄，也对橡树能够区分寒冷和真正的冬天来临感到好奇。柳树对橡树的话很敏感，在"看见"这个话题上，误认为橡树觉得他很蠢。误会解除后，柳树开始询问橡树没有眼睛怎么可能"看得见"，当橡树重点解释这一点时，柳树承认自己有点晕头转向。随着惊喜的出现，他的怀疑变成了笑声。现在，他放松了下来，享受着交谈的乐趣。他询问橡树，当橡树为其他生物提供食物的时候，是如何做到只允许被吃掉部分叶子。随着一系列奇妙的发现，柳树得出结论，橡树的规划确实为许多生物创造了有利于生命的条件。

橡 树

高寿的橡树有自己固有的思维。她不认为自己的年龄有什么特别之处，也不在意柳树的惊讶。当橡树意识到柳树非常敏感，并且自己的反应有可能伤及柳树的自尊时，她便耐心地向他解释自然界的多样性。这些事实对柳树来说是一个惊喜。橡树继续介绍自己的其他独特的特点，尽管这让柳树听得有些晕头转向。橡树保持耐心，讲述了细节，并解释她保护自己免受昆虫伤害的方式，以及避免在冬天被冻伤的方法，这显示出了她的同理心。通过机智和耐心，橡树成功地让柳树有了更积极的心态。

艺术
The Arts

画一棵橡树有很多方法，让我们来探索一下。访问HTTPS://FINE-ARTAMERICA.COM/ART/PAINTINGS/OAK+TREE，学习如何来描绘橡树。现在，请选择你想要的方式。展示你的画作，并解释你为什么要这么画。你的想象力是如何被激发的呢？

思维拓展
Systems: Making the Connections

　　100年前，橡树仍然被认为是一个重要的经济组成部分，但现在它的价值已经降到只能用作制造酿造威士忌和葡萄酒的高质量橡木桶。它作为建筑结构材料和造船材料的作用已经随着钢铁和水泥的出现而消失。过去，欧洲大教堂和日本寺庙的建筑师在完成他们划时代的杰作之后，会种植很多橡树，确保500年后能有足够的橡木去替换已达到使用寿命年限的横梁。这种习惯早已从今天大工业时代消失。在当今的工业标准中，供应链管理和准时交货制度认为这种延续几个世纪的习惯过于理想化了。打算几个世纪后取代梁木的橡树，在最近甚至受到法律的保护，丧失了橡树原本的功能和目的。种植和收获橡树需要耐心，因为一颗橡子从发芽到结出果实至少需要25年的时间，这迫使我们要在两代人之间进行长期规划。这对新一代来说是一项艰巨的任务，因为他们执着于立即能得到的满足感和在短时间内取得切实的成果。然而，对于橡树，不仅需要我们投入时间和耐心，也需要我们去探索，因为有太多关于橡树的奥秘，我们尚不知晓。我们不仅要学习更多关于橡树的知识，还要学习橡树的精神。橡树为我们提供了追求长寿、幸福和健康生活的洞见，只要我们准备好关注有记载以来就陪伴我们的这种非凡树种所提供的智慧。

动手能力
Capacity to Implement

　　让我们用延续了14个世纪的方法来制作墨水。先去森林里找一些橡树瘿。研究最佳的橡树瘿类型，并根据形状和大小进行选择。橡树瘿中的单宁酸可以用来制作防水并且不褪色的墨水。你需要三种原料：硫酸亚铁、阿拉伯树胶和一些橡树瘿。将两粒橡树瘿研碎，在半升水里浸泡一夜。再加入25毫升硫酸亚铁和12克阿拉伯胶。现在，你可以按照达芬奇的方式来创作你自己的杰作啦。

故事灵感来自
This Fable Is Inspired by

乔治·麦加文
George McGavin

乔治·麦加文在爱丁堡大学学习动物学，并获得伦敦帝国理工学院的昆虫学博士学位。他在牛津大学教书和做研究。他热衷于向公众传播科学。他为英国广播公司（BBC）和探索频道工作，还大胆地涉足烹饪和食用昆虫领域，他曾把昆虫称为"飞翔的大虾"。2017年，麦加文主持了BBC纪录片《橡树：大自然最伟大的幸存者》。此后，他制作了一部名为《垃圾填埋场的秘密生活：垃圾的历史》的纪录片。麦加文是一位不断用基于坚实科学的惊喜来激励观众的学者。

图书在版编目(CIP)数据

冈特生态童书.第七辑:全36册:汉英对照 /
(比)冈特·鲍利著;(哥伦)凯瑟琳娜·巴赫绘;
何家振等译.—上海:上海远东出版社,2020
ISBN 978-7-5476-1671-0

Ⅰ.①冈… Ⅱ.①冈…②凯…③何… Ⅲ.①生态
环境–环境保护–儿童读物—汉英 Ⅳ.①X171.1-49

中国版本图书馆CIP数据核字(2020)第236911号

策　　划	张　蓉
责任编辑	程云琦
封面设计	魏　来　李　廉

冈特生态童书

永葆生机

[比]冈特·鲍利　著
[哥伦]凯瑟琳娜·巴赫　绘
章里西　译

记得要和身边的小朋友分享环保知识哦!
八喜冰淇淋祝你成为环保小使者!